CHAIN REACTIONS

From Mendel's Peas to Genetic Fingerprinting

Discovering Inheritance

Sally Morgan

Heinemann Library
Chicago, Illinois

Customer Service 888-363-4266
Visit our website at www.heinemannlibrary.com

Consultant: Michael Reiss
Commissioning editor: Andrew Farrow
Editors: Kelly Davis and Richard Woodham
Proofreader: Catherine Clarke
Design: Tim Mayer
Picture research: Amy Sparks
Artwork: William Donohoe pp. 8, 11, 15, 16, 18, 24;
Wooden Ark pp. 31, 32.

Originated by RMW
Printed and bound in China by South China
Printing Company

10 09 08 07 06
10 9 8 7 6 5 4 3 2 1

Library of Congress Cataloging-in-Publication Data
Morgan, Sally.
  From Mendel's peas to genetic fingerprinting :
discovering inheritance /
Sally Morgan.
     p. cm. -- (Chain reactions)
  Includes bibliographical references and index.
  ISBN 1-4034-8837-1 (hc)
  1. Genetics--History--Juvenile literature. 2. Heredity--
History--Juvenile
literature. I. Title. II. Series.
  QH437.5.M64 2007
  576.509--dc22
                              2006011043

Acknowledgments. The author and publisher would like
to thank the following for allowing their pictures to be
reproduced in this publication:

Corbis pp. 21 (Frank Trapper), 22 (Bettmann),
26 (Bettmann), 34 (Bettmann), 39 (Bettmann), 52 (Karen
Kasmauski); Ecoscene p. 25 (Sally Morgan); Index Stock
Imagery pp. 1 (Oxford Scientific Films Photolibrary),
9, 23 (Oxford Scientific Films Photolibrary); Science Photo
Library pp. 4 (Peter Menzel), 5 (Michael Donne), 6 (James
King-Holmes), 7 (Dr. Jeremy Burgess), 12, 13 (Herve
Conge, ISM), 14 (Dr. Bernard Lunaud), 16, 17 (Biophoto
Associates), 19, 20 (Eye of Science), 27 (Biology Media),
28, 30 (A. Barrington Brown), 33 (National Library of
Medicine), 35 (Peter Menzel), 36 (Robert Longuehaye,
NIBSC), 37 (Peter Menzel), 38 (Simon Fraser/RVI,
Newcastle-upon-Tyne), 40 (Dr. Tony Brain), 41 (Sam
Ogden), 42 (David Parker), 43 (James King-Holmes),
44 (Volker Steger), 45 (David Parker), 46 (David Parker),
47 (Tek Image), 48 (Peter Menzel), 50 (James King-
Holmes), 51 (CNRI), 54 (Victor de Schwanberg), cover
(Coneyl Jay); Topfoto.co.uk pp. 29, 55 (IMW).

Cover design by Tim Mayer.

The paper used to print this book comes from
sustainable resources.

Disclaimer
All the Internet addresses (URLs) given in this book were
valid at the time of going to press. However, due to the
dynamic nature of the Internet, some addresses may
have changed, or sites may have ceased to exist since
publication. While the author and publishers regret any
inconvenience this may cause readers, no responsibility
for any such changes can be accepted by either the
author or the publishers.

Contents

Any words appearing in the text in bold, **like this**, are explained in the Glossary.

From Growing Pea Plants to Identifying Criminals

Imagine that somebody has been attacked. Forensic scientists are searching for clues to help them identify the person who committed the crime. They are looking for fingerprints, but not the ones left by fingertips. They are searching for something that will give them a source of **genetic** information, like a drop of blood, a sliver of skin, or the root of a hair. Back in the lab, they will extract the **DNA (deoxyribonucleic acid)**, and produce a genetic fingerprint.

Every person has a unique genetic fingerprint. The forensic scientists will compare the genetic fingerprint from the crime scene to others on their database. If all goes well, they will identify the criminal. Forensic science is just one field that is making use of advances in our understanding of how characteristics are passed from one generation to another. This area of study is known as genetic inheritance.

A forensic scientist takes a sample of dried blood from a shirt during a murder investigation. The blood will be analyzed to provide evidence about the people involved in the crime.

The story of inheritance

Have you ever noticed the strong similarity between parents and their children? The study of inheritance, or heredity as it is sometimes called, investigates the reasons for the similarities and differences between individuals.

The story of inheritance goes back about 150 years to an Austrian monk named Gregor Mendel (1822–1884), who was fascinated by plants. He carried out thousands of experiments on pea plants. The results enabled him to come up with a theory that explained how certain features of the plants were inherited.

As microscopes improved scientists were able to look at cells in greater detail. Soon they discovered structures called chromosomes within the **nucleus** (control center) of a cell. Chromosomes contain DNA.

For a long time, scientists struggled to work out how the DNA molecule was constructed. Two researchers, Francis Crick (1916–2004) and James Watson, finally worked out its structure in 1953. This was a major breakthrough. Since then scientists have learned how to alter DNA. They have worked out the complete genetic code of human DNA. They have even traced the way humans have evolved over hundreds of thousands of years by comparing the DNA of several different ethnic groups.

In this book you will learn about the remarkable chain of events that began with growing garden peas and led to **genetic fingerprinting**. You will read about the many successes and failures that occurred along the way. You will also find out about the challenges that remain for scientists of the future.

TALKING SCIENCE

"DNA typing [fingerprinting] has been called the most important discovery in forensic science since fingerprints." Dale Laux, Forensic Scientist, Ohio Bureau of Criminal Investigation, 2002

Forensic scientists inspect the shallow grave of a dummy representing a murder victim. They wear protective clothing and masks to avoid contaminating the crime scene with their own DNA.

Proving Inheritance

The first steps in the understanding of inheritance date back to the time of Gregor Mendel, who lived in the 1800s. He was a monk who worked as a teacher in a monastery in Brno, in what is now the Czech Republic. His research with pea plants provided the basis for the study of inheritance. His breakthrough was all the more amazing because his studies took place long before the discovery of chromosomes and **genes**.

Mendel had read the work of Jean Baptiste Lamarck, a French naturalist. Lamarck thought that living things changed their behavior in response to changes in their environment. For example, Lamarck believed that a giraffe gained its extra-long neck and front legs because it had to stretch up to reach leaves on high branches. Over time, this lengthened its neck and legs. Its offspring then inherited these characteristics. Mendel decided to investigate how inheritance worked.

The Austrian monk and botanist Gregor Mendel grew his garden peas in the monastery gardens at Brno. Mendel recorded the results of his experiments in great detail. His conclusions provided the basis for the study of **genetics** today.

Working with plants

During the 1850s Mendel started a series of experiments with garden peas. He noticed that the pea plants had different characteristics. For example, some seeds were wrinkled and others were smooth. Some had purple flowers and others had white ones. He decided to study seven of these characteristics:

1. purple or white flower color

2. flowers at the top of the stem or on the side of the stem

3. inflated or constricted seed pod

4. yellow or green pod color

5. yellow or green seed color

6. round or wrinkled seeds

7. long or short stems.

He grew many pea plants and then chose two of them to study. He collected seeds from these two parent plants. Then he grew a new generation of pea plants. He counted how many of the plants had features of one parent, and how many had features of the other.

The garden pea flower is usually pollinated by insects. The insect crawls into the middle of the flower, where pollen rubs onto its body. When the insect flies to another flower, the pollen is transferred.

? HOW DID MENDEL POLLINATE THE PLANTS?

Mendel had to make sure that the flowers of his parent plants were not pollinated by other plants. To prevent this, he covered the flowers, and carried out the pollination himself. First, he removed the **stamens** (the male parts) from the flowers that were going to produce seeds. This made it impossible for them to pollinate themselves. Then, he transferred pollen onto their **stigmas** (female parts). A few weeks later, he collected the seeds and started growing them.

Inheriting characteristics

Mendel's first experiments showed that a pea plant's offspring kept its parents' characteristics. This disproved Lamarck's ideas about living things being changed by their environment.

For example, Mendel looked at the inheritance of yellow and green seeds. He crossed a pea plant that produced yellow seeds with a pea plant that produced green seeds. He collected the seeds and germinated them. These first-generation seeds grew into plants that all had yellow seeds. Then he crossed two of these plants to produce the second generation. In the second generation, three-quarters of the plants had yellow seeds and one-quarter had green seeds.

This diagram shows the results of Mendel's experiment with yellow and green seeds.

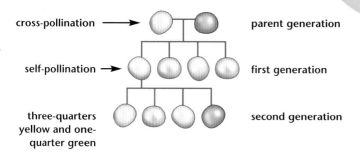

cross-pollination ➝   parent generation

self-pollination ➝   first generation

three-quarters yellow and one-quarter green   second generation

Dominant or recessive?

When the plants were crossed there was no blending of the colors. The plants had either green seeds or yellow seeds, and no greenish-yellow seeds. Mendel therefore concluded that one characteristic must be **dominant**, and the other must be **recessive**. A dominant characteristic will mask a recessive one. For example, do you have ear lobes? This is a dominant characteristic. If you do not have ear lobes, you are recessive for this characteristic. In Mendel's experiments, there were no green seeds in the first generation of plants because yellow was the dominant color in the parent generation. The yellow masked the presence of the green color.

Mendel's laws

To make this happen, Mendel knew that something he called a **factor** must be transmitted from parent to offspring. But how was this factor transmitted? He thought the only way was through the **gametes**.

Mendel also concluded that each parent had a pair of factors for each characteristic. But only one of these two factors was present in the gamete. When the gametes fused, the offspring ended up with one factor from each parent. He called this the law of segregation.

Peas are easy plants to grow, and they make a lot of seeds.

He also suggested that each factor was inherited separately from other factors. For example, a pea plant might inherit the factor for purple flowers. However, that did not mean it would also inherit the factor for yellow peas rather than green ones. This became known as the law of independent assortment.

Lucky choice

Mendel's decision to use garden peas was very lucky because the seven characteristics that he selected gave clear results. Pea flowers are also easy to pollinate and they produce many seeds. He grew thousands of plants and this meant that his results were more reliable. By the time his experiments were completed, he had examined about 10,000 plants.

In 1866 Mendel published his findings in a paper called "Experiments in Plant Hybridization" in the *Proceedings of the Natural Science Society of Brno*. His paper was brilliant, but other scientists did not recognize its importance.

Mendel carried on with his research, but this time using a plant called the milkweed. This plant behaved very differently from the garden pea. Mendel was disappointed by the results he got with milkweed, and gave up his studies. He eventually became the abbot of his monastery at Brno, where he died in 1884. For 35 years his research paper was forgotten until it was suddenly rediscovered in 1900 (see page 14).

What are Mendel's factors?

We now know that the factors described by Mendel are genes. A gene controls a particular characteristic, such as flower color, or whether or not a person has ear lobes. Genes exist in different forms. These forms are called **alleles**. For example, the gene controlling flower color in peas comes in two forms, one for purple and one for white. The purple allele is dominant, while the white one is recessive.

An individual will have two alleles for each characteristic. If the alleles are the same, the individual is described as **homozygous**. If the alleles are different, the individual is called **heterozygous**. However, there is only one allele for each gene in the gamete. This means that an individual inherits one allele from each parent.

WHAT HAPPENS WHEN YOU CROSS PURPLE FLOWERS WITH WHITE FLOWERS?

Imagine crossing a purple-flowered pea plant with a white-flowered pea plant. You know that both parents are homozygous for flower color. This means that the alleles are the same. The alleles can be represented by letters (a capital P for purple, and a small italic *p* for white). In the experiment below, all the first generation of pea plants have purple flowers. But they are not homozygous because they have one dominant allele and one recessive allele. When two of these pea plants are crossed, they produce purple plants and white plants.

Key:
P = purple flower allele
p = white flower allele

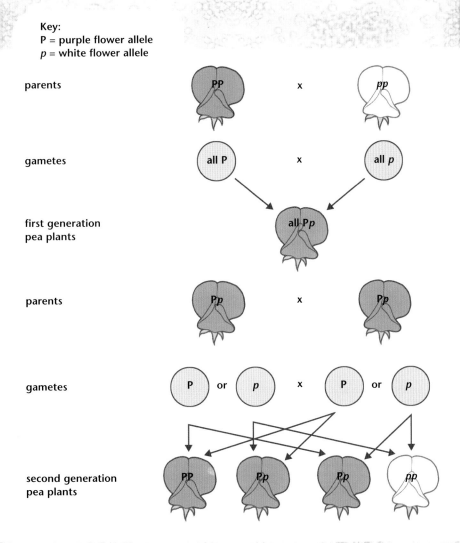

Discovering Chromosomes

During the 1880s improved microscopes enabled scientists to see cells in much more detail. This increased the scientists' interest in the structure and development of cells.

Theodor Boveri was a brilliant scientist. In 1914, towards the end of his life, he made a very far-sighted suggestion. He thought that abnormal chromosomes might be responsible for cancerous tumors. We now know that this is partly true.

Looking at cells

Chromosomes are found in the **nucleus** of a cell. Chromosomes were first observed in plant cells in 1842 by the Swiss botanist Karl Wilhelm von Nägeli (1817–1891). However, von Nägeli did not know what they did in the cell. No further progress was made until the 1880s.

Walter Flemming (1843–1905), a German scientist, was interested in the way cells divide to form two new cells. However, he found it difficult to see what was happening in the cell's nucleus. He produced some dyes that were absorbed by the structures in the nucleus. These dyes made the structures stand out more clearly when viewed under a microscope. Flemming noticed that long, thin threads appeared in the nucleus just before the cell started to divide. He carefully recorded the behavior of these threads during cell division. In 1882 he published his findings in a book. These thread-like structures were later called chromosomes.

In 1887 Theodor Boveri (1862–1915), a German zoology professor, started some exciting research. He had been inspired by Walter Flemming's description of chromosomes. He suspected that chromosomes could perhaps be involved with inheritance.

Watching cells divide

For the next three years, Boveri studied cells dividing under a microscope. His favorite subject was a roundworm's egg. This contained only two chromosomes so it was easy to study. In 1888 Boveri published his results. He reported that just before cell division the chromosomes disappeared. But after cell division was complete, the chromosomes reappeared in both daughter cells. This suggested that chromosomes could be involved in inheritance.

Boveri then went on to study sea urchin eggs. He discovered that during fertilization the sperm and egg each contributed the same number of chromosomes. He concluded that a new individual is given one set of chromosomes from the mother, via the egg. It also receives a set of chromosomes from the father, via the sperm.

His results were published in 1890, and they attracted a lot of attention. But many scientists were still not convinced that chromosomes played a central role in inheritance.

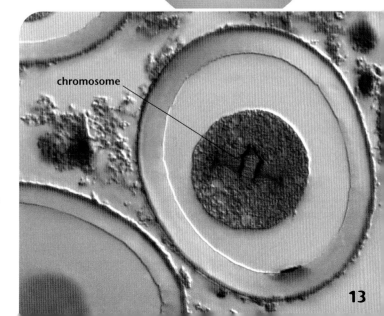

This roundworm egg cell, photographed under a microscope, is in the process of dividing. It has been magnified about 1,400 times.

chromosome

Chromosome roles

Next, Boveri looked at individual chromosomes. He wondered whether a full set of chromosomes was needed for healthy development of an organism. In his experiments with sea urchin eggs, Boveri showed that each individual chromosome had a specific role in development. If a chromosome was missing, or there were two of the same chromosome, the organism did not develop correctly.

By now Mendel's work had been rediscovered. Boveri saw that Mendel's laws of segregation and assortment (see page 9) could apply to chromosomes. He believed that the chromosomes contained the **factors** mentioned by Mendel in his paper. Today these factors are known as **genes**. In 1903, Boveri concluded that the factors in Mendel's experiments were connected to specific chromosomes.

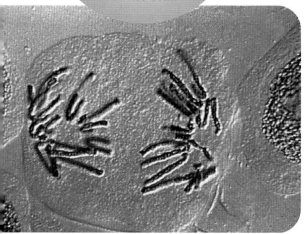

This photograph, taken under a microscope, shows a bluebell plant cell in the later stages of mitosis. The orange, thread-like chromosomes have split into two, and are being pulled to either end of the cell.

The Boveri-Sutton chromosome theory

Walter Sutton (1877–1916) was an American scientist studying chromosomes at the same time as Boveri. He had been studying the type of cell division that is responsible for making the sex cells known as **gametes**. This is called **meiosis** (see page 15).

WHAT IS MITOSIS?

In **mitosis** a cell divides into two identical daughter cells. The chromosomes copy themselves. Then the two copies of each chromosome are separated. Mitosis is used in growth.

Gametes are different from other cells because they contain half the normal number of chromosomes. Humans have 46 chromosomes in most of their cells. However, eggs and sperm have only 23 chromosomes each. When the egg and sperm fuse, the **embryo** therefore has 46 chromosomes. If the number were not halved, the embryo would have 92 chromosomes.

In 1902 Sutton suggested that chromosomes occurred in pairs. He also suggested that in meiosis the pairs of chromosomes were carefully separated. This is how the gamete ends up with one of each chromosome. In 1903, he published his work under the title "The Chromosomes in Heredity."

The Boveri-Sutton chromosome theory was discussed by scientists around the world for several years. Many scientists rejected the theory completely. However, the matter was settled a few years later, when Thomas Hunt Morgan published his work on the fruit fly (see page 25) and proved them right.

? WHAT IS MEIOSIS?

In meiosis (or a reduction division) the number of chromosomes is halved to form the gametes. Meiosis starts with the chromosomes copying themselves in the nucleus. The pairs of chromosomes move apart, and the first division separates the pairs. In the second division, the copies of each chromosome are separated. Then the nuclei split. The result is four cells, with each cell containing half the number of chromosomes.

1. The nucleus contains four chromosomes.

2. Each chromosome copies itself.

3. The chromosome pairs move apart.

4. The nucleus divides.

5. The second division separates the copy strands of each chromosome.

6. Four nuclei are formed, each with half the number of chromosomes contained in the parent cell.

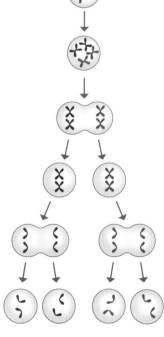

Studying sea urchin eggs

Once chromosomes had been discovered in cells, scientists started to investigate further. Two American researchers, Nettie Stevens (1861–1912) and Edmund Beecher Wilson (1856–1939), worked independently on the idea that an animal's gender was determined by a chromosome.

Nettie Stevens discovered that the sex of an insect was determined by a sex chromosome present in the gametes. Her early death meant that her work was developed by others.

Edmund Beecher Wilson studied sea urchin eggs at Columbia University in New York during the 1880s and 1890s. He exchanged ideas about chromosomes with his friend Theodor Boveri. Wilson's studies made Boveri think that chromosomes contained **genetic** material.

Sex chromosomes

In 1902 Clarence McClung, a professor at Kansas University, identified a structure in a grasshopper's cell nucleus. He called this an "accessory chromosome." However, when he examined sperm from the grasshopper he could only find the accessory chromosome in half the sperm. He thought this chromosome must be involved in determining sex.

In 1905 Wilson counted and identified the chromosomes of sperm cells before and after meiosis. In two species of insect, the female insect had one more chromosome than the male. He decided that this extra chromosome was linked to an insect's sex.

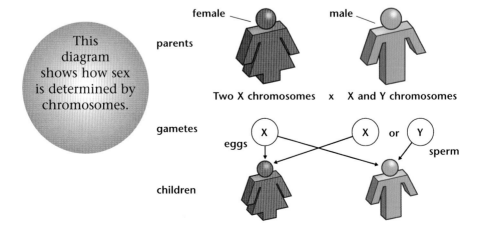

This diagram shows how sex is determined by chromosomes.

parents — female — male
Two X chromosomes x X and Y chromosomes
gametes — eggs — X x X or Y — sperm
children

At the same time, Nettie Stevens was carrying out similar studies on a different species of insect. She found that the male had a small Y-shaped chromosome that went into half the sperm. There was also a larger chromosome that went into the rest of the sperm. The female insect had two of the larger chromosomes. Like Wilson, Stevens concluded that the sex of an individual was determined by the presence of these chromosomes.

Stevens' and Wilson's findings, published within days of each other in 1905, formed the basis for the modern understanding of sex determination. They had proved that there were chromosomes that were responsible for determining sex. Years later their research would also help Thomas Hunt Morgan understand the results of his experiments with the fruit fly (see pages 23–24).

HOW ARE SEX CHROMOSOMES ARRANGED?

Humans have 46 chromosomes. These are arranged in 22 pairs of **autosomes** (not linked to sex), and one pair of sex chromosomes. The sex chromosomes are known as X and Y. There are two X chromosomes in females, and one X and one Y chromosome in males. The Y is smaller and does not carry as many genes as an X chromosome. All egg cells contain one X chromosome. Half the sperm cells carry an X chromosome, and the other half carry a Y chromosome. This means there is always a 50–50 chance that a baby will be male or female.

These are the 46 chromosomes from a healthy human female. Each cell contains 22 matched pairs, and one pair of sex chromosomes.

Genetic Diseases

Humans suffer from many diseases. Some of these diseases, such as cystic fibrosis and sickle cell anemia, are caused by **genetic** defects. The genetic material is somehow damaged or altered, and this causes a disease. The link between inheritance and disease was first investigated by Sir Archibald Garrod (1857–1936), an English doctor, in the early 1900s.

Inheritance and the workings of cells

Garrod was interested in a disease called alkaptonuria. One of the few symptoms of this disease is urine turning black when it is exposed to air. Garrod collected family histories and urine samples from his patients. He noticed that the disease occurred in certain families, and concluded that the disease must be inherited. He knew about Mendel's work and he decided that alkaptonuria was a **recessive** disease. This meant that a person could inherit the disease even if neither parent suffered from it. Garrod published his findings in 1902.

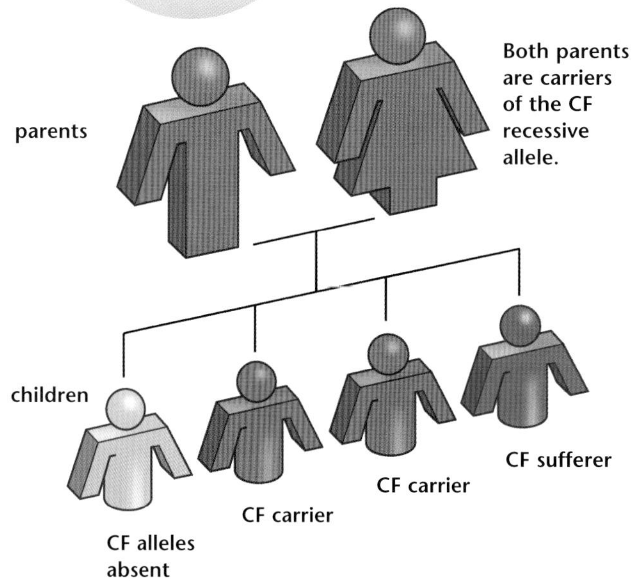

Garrod also thought that alkaptonuria was caused by a malfunction that resulted in an abnormal substance being released in the urine. Scientists at that time knew little about how the body dealt with waste substances. However, we now know that alkaptonuria is caused by an inability to break down certain **amino acids** (which together make up proteins). This results in a chemical building up in the skin and other tissues. Some of this chemical goes into the urine, causing it to turn black when left in air.

This diagram shows how cystic fibrosis (CF) is inherited.

parents

Both parents are carriers of the CF recessive allele.

children

CF alleles absent

CF carrier

CF carrier

CF sufferer

Garrod went on to study other inherited conditions, including albinism. A person with albinism cannot produce any pigment in their skin, hair, or eyes. As a result, they have pale skin, white hair, and pink eyes. It is another recessive condition. Garrod's work became the basis for later research into how **genes** can control a cell's workings.

Sir Archibald Edward Garrod used family histories to show that certain diseases were inherited.

HOW ARE GENETIC DISEASES INHERITED?

Most genetic diseases are caused by **recessive alleles**. In order to get the disease, a person must inherit a recessive **allele** from both their parents. Their parents may be perfectly healthy, but still carry a recessive allele. This allele is masked by the healthy **dominant allele**. They are therefore unaware that they are carrying the genetic disease. For example, in North America, about 1 in 25 people carry the faulty gene for cystic fibrosis and one in 2,500 children are born with the disease. If two carriers have a child, there is a 25 percent chance that the child will suffer from cystic fibrosis.

Tay-Sachs disease and sickle cell anemia

Since Garrod's time, many other inherited diseases have been discovered. Scientists have used Garrod's work to help people with these inherited disorders understand them, and make decisions about having children.

None of these diseases can be cured, but sometimes the symptoms can be eased. For example, some patients may be given special diets. However, most inherited disorders are impossible to control. Tay-Sachs disease is a genetic disorder. Babies born with this disease appear normal at birth. But they gradually lose control of their muscles, and they die at about three years of age. Sadly there is no treatment at present.

Sickle cell anemia is another inherited genetic disease that is caused by a recessive allele. The faulty allele changes the structure of hemoglobin, a protein found in red blood cells. Hemoglobin picks up oxygen as the red blood cells flow through the lungs. The faulty hemoglobin does not pick up as much oxygen. The person may therefore get breathless and tired, and suffer a great deal of pain.

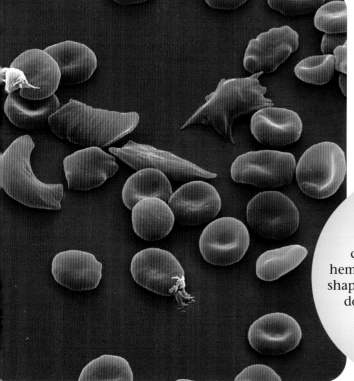

In a person with sickle cell anemia, the red blood cells containing abnormal hemoglobin have a crescent shape, instead of the normal doughnut shape. These cells have been magnified about 3,000 times.

Dominant diseases

There are a few diseases caused by dominant alleles. Two examples are achondroplasia and Huntington's disease, also called Huntington's chorea. Achondroplasia causes dwarfism, a condition in which the bones do not grow properly, resulting in a stunted body and bow legs.

Huntington's disease affects nerve cells in the brain, and gets progressively worse. The symptoms include loss of muscle control, depression, and memory loss. However, these symptoms do not usually appear until the person is between 30 and 50 years old. By this time they may have had a family and passed the gene onto their children. The disease was first described by George Huntington, an American doctor, more than 100 years ago. But he did not know the cause. We now know that the disease is caused by a faulty gene that results in damage to nerve cells in the brain.

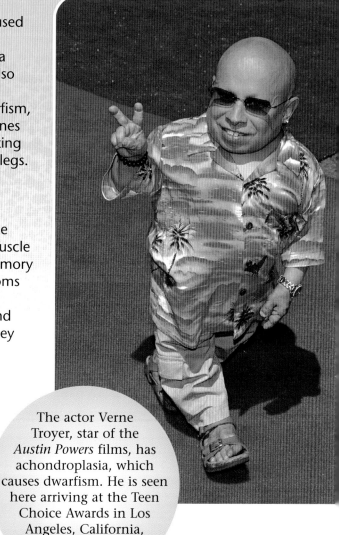

The actor Verne Troyer, star of the *Austin Powers* films, has achondroplasia, which causes dwarfism. He is seen here arriving at the Teen Choice Awards in Los Angeles, California, in 2002.

? WHAT IS HUNTINGTON'S DISEASE?

Huntington's disease has been known at least since the Middle Ages. One of the earliest names for it was "chorea," the Greek word for "dance" (as in "choreography"). The word "chorea" refers to the tendency of sufferers to twist and turn in a constant, uncontrollable motion, as if they were dancing. Anyone who has a parent with Huntington's disease has a 50 percent chance of inheriting the faulty gene.

Working with Fruit Flies

When Wilson and Stevens published their findings on the sex chromosomes of insects in 1905 (see page 17), several scientists doubted the results. One of them was Thomas Hunt Morgan (1866–1945). Although he admitted that the chromosomes might have something to do with inheritance, he did not believe that a chromosome could carry specific **genes**.

The geneticist Thomas Hunt Morgan is seen here with a microscope in his laboratory. This photograph was taken in the mid-1920s.

In 1904 Morgan moved to Columbia University to become a professor of experimental zoology. He had started to study inheritance in mice and rats. However, he found that they reproduced too slowly, so he switched to the fruit fly (*Drosophila melanogaster*). These flies bred so quickly that Morgan could study 30 generations in a single year. In addition the fruit flies only had four pairs of chromosomes.

The key breakthrough came in 1910 when Morgan discovered a male fly with white eyes rather than red eyes. This change in appearance was due to a mutation. Mutations are sudden changes to **genetic** material. Sometimes the changes can be major ones and the organism dies. At other times the mutations are relatively minor, such as white eyes rather than red. Most genetic diseases are also caused by mutations.

Morgan had three main questions:

- Where did this change in eye color come from?

- What determines eye color?

- Could eye color be inherited?

His experiments showed that the red eye color was **dominant** and the white eye color was **recessive**. But he then found out that only male flies had white eyes, and all the females had red eyes. This must mean that the inheritance of eye color was linked to the inheritance of sex. The genes for eye color and sex therefore had to be located on a sex chromosome. In time other mutations appeared. These were tiny wings and yellow body color. These mutations were also linked to the sex of the fly.

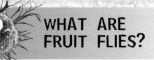

WHAT ARE FRUIT FLIES?

The fruit fly gets its name because of the way the flies gather and feed on fruit. This type of fly is very small and very easy to breed in the laboratory. As many as 1,000 flies can be kept in a 2 pint (1.1 liter) bottle. The females lay many eggs, which hatch quickly and turn into **larvae**. Then they **pupate** and emerge as adults just 12 days later.

Fruit flies are 3 millimeters long. They are used in genetic research because they only have four pairs of large chromosomes. This makes the fruit fly easy to study and identify.

Genes and sex chromosomes

When Morgan looked at the fruit fly chromosomes under a microscope, he discovered that each pair of chromosomes was identical, except for the sex chromosomes in the male. This was similar to Wilson's and Stevens' findings (see pages 16–17). The female flies had two X chromosomes that looked identical. But in the male the X chromosome was paired with a Y chromosome. This Y chromosome looked different and was never found in the female.

Once Morgan had worked this out, he could see how the gene for eye color was inherited. A male fly inherits the X chromosome from its mother and the Y chromosome from its father. The gene for eye color is found only on the X chromosome. When the female fly has two **alleles** for red eyes, all her male offspring will have red eyes. This is because they inherit one of her X chromosomes. A male fly with white eyes can only be produced from a female with white eyes.

This diagram shows some of the results of Morgan's experiment with red and white eyed fruit flies.

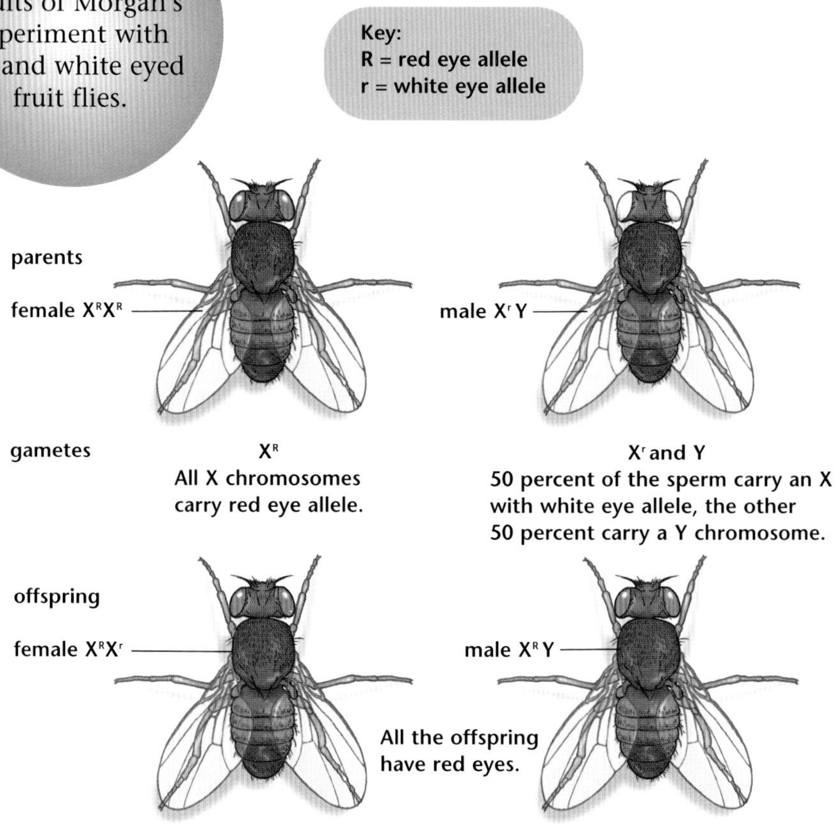

Key:
R = red eye allele
r = white eye allele

parents

female $X^R X^R$

male $X^r Y$

gametes

X^R
All X chromosomes carry red eye allele.

X^r and Y
50 percent of the sperm carry an X with white eye allele, the other 50 percent carry a Y chromosome.

offspring

female $X^R X^r$

male $X^R Y$

All the offspring have red eyes.

When he published his results in 1910, Morgan drew three conclusions:

● genes are found on chromosomes

● each gene must be located on a particular chromosome

● the gene for eye color is found on the X chromosome.

Linked genes

Thomas Morgan suggested that each chromosome was a series of small units called genes. Since certain genes were found together on one chromosome, they tended to be inherited together. He said that the genes were linked. For example, he found that the characteristics of long wing and broad abdomen were seen together. These linked genes would prove to be very useful when scientists started to map the positions of all the genes on chromosomes.

WHY ARE MORE MEN THAN WOMEN COLOR BLIND?

A person is "red-green color blind" when they cannot see the difference between red and green. Look at the pattern of dots in the photograph. People with normal color vision will see the number "29" but a color-blind person will just see a random pattern of dots. The reason why more men than women are color blind is that the gene for color blindness lies on the X chromosome. A woman has two X chromosomes so the recessive color-blindness allele can be masked by the allele for full-color vision. A man only has one X chromosome, so if he has the **recessive allele** it cannot be masked.

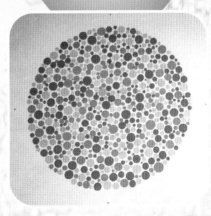

This card tests red-green color blindness. A person with normal vision will be able to see the number 29.

Mapping chromosomes

Alfred Henry Sturtevant was a 19-year-old student working with Morgan. He realized that he could use linked genes to map the position of genes on the chromosome. He decided that genes that were always inherited together must lie near each other on the chromosome. In 1911 Sturtevant drew up the first genetic map for the sex-linked genes. Morgan described the mapping of genes on the chromosome as "one of the most amazing developments in the history of biology."

Twenty years later more detail could be added to the genetic map. Sturtevant had only been able to give the order of the genes. A scientist named Calvin Bridges developed a map showing the exact physical location of a gene on a chromosome.

Dr. Calvin Bridges compares a magnified fly chromosome with a chart showing the locations of the genes on the fly's chromosomes.

Again, it was the fruit fly that provided the breakthrough. The fruit fly maggot has gigantic chromosomes in its salivary glands. When Bridges looked at them under a microscope, he could see a pattern of bands. Each band corresponded to the position of one gene. Bridges identified 1,024 bands on the X chromosome.

THAT'S AMAZING!

Thomas Hunt Morgan's laboratory was nicknamed the "Fly Room." This small room, just 16 x 23 feet (5 x 7 meters), was used by all the researchers on the fruit fly project. The students carried out their own experiments, but they told each other about their results. This enabled them to develop new experimental techniques. Their way of working together was copied in other science laboratories around the world. Morgan's students went on to make amazing discoveries of their own after they left the Fly Room. Five of them went on to win Nobel Prizes.

Questions for the future

Back in the early 1900s, Thomas Hunt Morgan's experiments raised several questions:

● What is the chemical nature of genes?

● How do genes copy themselves?

● What goes wrong when genes mutate?

● How do genes help us understand genetic disease?

● How do genes control the development of organisms?

● What is the role of genes in evolution?

Some of these questions were answered by Morgan and his team, but others would not be answered for many decades. Morgan's pioneering research influenced the study of biology, especially genetics, for the rest of the century.

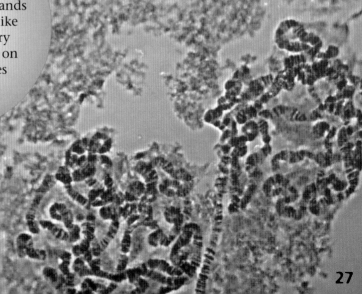

This highly magnified photograph of the fruit fly's salivary glands shows the thread-like chromosomes very clearly. Each band on the chromosomes corresponds to a gene.

The DNA Story

Thomas Hunt Morgan had proved that the **genes** on chromosomes carried **genetic** material. Now scientists had to find the chemical from which the genes were made.

The first clue came in 1869 when a Swiss doctor, Johann Friedrich Miescher (1844–1895), discovered a large molecule in the **nucleus** of cells, which he called "nuclein." He suggested that nuclein could be involved with inheritance. Sixteen years later, Edmund Beecher Wilson found a chemical called nucleic acid in the nucleus. We now know that there are two nucleic acids: **DNA (deoxyribonucleic acid)** and **RNA (ribonucleic acid)**.

Johann Friedrich Miescher, the discoverer of DNA, became a professor of physiology at Basel, Switzerland, in 1871.

The next step forward came in 1924, when scientists looked at cells under microscopes and used dyes to stain certain parts of the cells. These studies showed that chromosomes were made of DNA and protein. However, most scientists felt that DNA was unlikely to carry genetic information. They thought protein was a more likely candidate.

Transforming bacteria

Were genes made of DNA or protein? The final proof was obtained using **bacteria** (simple, single-celled organisms). Some bacteria cause disease. The bacterium named *Pneumococcus* exists in two forms: S strain and R strain. S strain is harmful. However, R strain is completely harmless.

In 1928 Frederick Griffith, a scientist in London, England, managed to change the harmless R strain into the harmful S strain. He did this by adding dead S-strain bacteria to a liquid in which the R-strain bacteria were growing. He thought an unknown chemical must have somehow passed from the dead bacteria to the R-strain bacteria.

This unknown chemical was not identified until 1944. Three American scientists, Oswald Avery, Colin MacLeod, and Maclyn McCarty, carried out some more experiments using the same type of bacteria and proved that the chemical contained DNA. Suddenly scientists around the world became interested in DNA.

WHAT IS X-RAY DIFFRACTION?

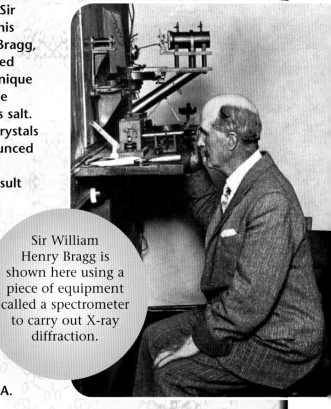

In 1912 a British physicist, Sir William Henry Bragg, and his son, Sir William Lawrence Bragg, developed a technique called X-ray diffraction. This technique could be used to look at the structure of crystals such as salt. They bombarded the salt crystals with X-rays. The X-rays bounced off the crystals and onto a photographic plate. The result was a pattern of dots. By studying the pattern of dots, scientists could work out the structure of the crystal. During the 1950s a group of scientists led by Maurice Wilkins at King's College, London, used X-ray diffraction to look at the structure of DNA.

Sir William Henry Bragg is shown here using a piece of equipment called a spectrometer to carry out X-ray diffraction.

Four bases

During the 1930s and 1940s, scientists found that DNA was made up of several different molecules. It contained sugars, phosphates, and four molecules known as bases: adenine, thymine, cytosine, and guanine. But nobody knew how these parts were put together to make DNA.

In 1949 a scientist named Erwin Chargaff found that the amount of adenine in DNA always equaled the amount of thymine. The amount of guanine was also always the same as the amount of cytosine. It would be four years before the significance of his results would be understood.

Putting it all together

James Watson was an American working at Cambridge University, in the United Kingdom. He had attended a lecture by Maurice Wilkins on DNA and X-ray diffraction. Now he was keen to solve the mystery of the structure of DNA. He worked on this with an Englishman named Francis Crick. Watson and Crick studied all the research that had been carried out by others in the field. They also built models of the DNA molecule and tried to fit all the parts together.

Meanwhile, Maurice Wilkins and Rosalind Franklin (a member of Wilkins' team at King's College) were using X-ray diffraction to look at DNA. Using this method they had worked out that DNA was a spiral or twisted molecule, known as a helix.

Watson and Crick looked at Franklin's X-rays, and decided that there must be two parallel twisted chains: a double helix. From the position of the spots on the X-rays, they thought the DNA molecule must look like a twisted ladder. The sides of the ladder were made up of alternating sugars and phosphates.

James Watson (left) and Francis Crick (right) are shown with their model of a DNA molecule at the Cavendish Laboratory, in Cambridge.

Rosalind Franklin worked with Maurice Wilkins during the 1950s, carrying out X-ray diffraction on samples of pure DNA. One of her high-quality X-rays enabled Crick and Watson to work out the structure of DNA. Sadly Rosalind Franklin died of cancer a few years later. In 1962, Crick, Watson, and Wilkins were awarded a Nobel Prize for their work on DNA.

Finally, Watson and Crick had to work out how the bases were arranged. From Chargaff's work they knew that there were equal amounts of adenine and thymine, and equal amounts of guanine and cytosine. Watson realized that if you joined adenine and thymine together, and guanine and cytosine together, the pairs of bases produced "rungs" of the same length.

In 1953 Watson and Crick announced their findings. Now scientists understood the structure of DNA, they could work out how the information carried by the DNA could be copied. They could also find out how the DNA instructed the cell to make proteins, and how DNA differed between species.

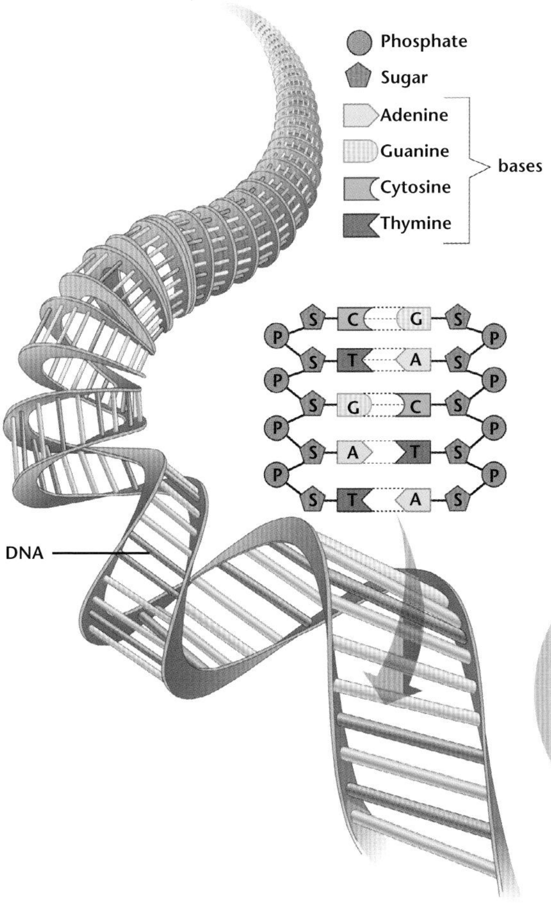

Phosphate
Sugar
Adenine
Guanine
Cytosine
Thymine
bases

DNA

This diagram shows the double helix structure of DNA, with its rungs made from pairs of bases.

Cracking the code

DNA contained all the genetic information that was passed from generation to generation, but how was it stored? The only possible place was in the sequence of the four bases, along the length of the DNA molecule. Scientists suspected that the genetic information was in the form of a code.

The DNA making up one chromosome may have many hundreds of genes along its length. Each gene contains information on how to build a particular protein. The information is like a recipe. It states the type and number of **amino acids** needed to make a particular protein. It also states the order in which they have to be joined together.

HOW DO AMINO ACIDS FORM PROTEINS?

A protein is a large molecule, made up of amino acids. The amino acids are joined together in a chain by bonds. The chain then twists around to form a strand or a ball-shaped protein. The shape of a protein is important because its role is linked to its shape. The shape is determined by the sequence of amino acids.

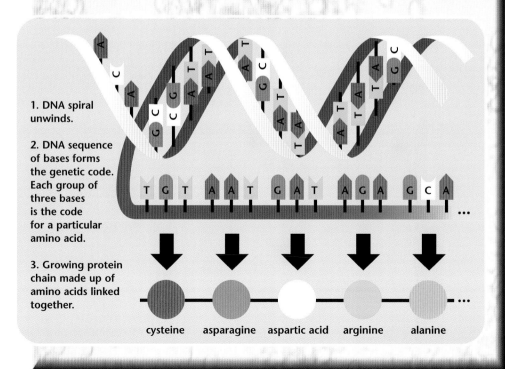

1. DNA spiral unwinds.

2. DNA sequence of bases forms the genetic code. Each group of three bases is the code for a particular amino acid.

3. Growing protein chain made up of amino acids linked together.

cysteine   asparagine   aspartic acid   arginine   alanine

Marshall Nirenberg is seen here at work in his laboratory. He shared the 1968 Nobel Prize in Physiology or Medicine for his work on genetics.

Every protein in the human body is made from just 20 amino acids. These amino acids can be joined in different combinations, and in different orders. It is vitally important that the amino acids are joined in the correct order, otherwise the protein may not work properly. The amino acid sequence is directly linked to the sequence of bases on the DNA molecule.

The information in the DNA molecule is in the form of a coded message. Scientists knew that the code could only be made up of four parts, as there are only four bases in DNA, represented by the letters A, C, G, and T.

During the early 1960s, two groups of scientists were trying to crack the code. Francis Crick and Leslie Barnett were working in the United Kingdom, and Marshall Nirenberg and Heinrich Matthaei were working in the United States. Crick and Barnett worked out that three bases were needed to code for a single amino acid.

By 1966 Nirenberg and Matthaei had worked out all the individual codes. With combinations of three letters, it was possible to have 64 different codes, for example AAA, ACG, GCC, and GGG. Since there were only 20 amino acids, there were plenty of spare codes. In fact they discovered that some amino acids had more than one code. For example, both AAG and AAA code for the amino acid lysine. Nirenberg and Matthaei also found a combination that signaled the start of a code, and another that indicated the end of the code.

Reading the code

Now scientists understood the code, they wanted to work out the complete sequence of bases that make up a gene. But how could they separate the piece of DNA that they wanted to study? And how could they copy the DNA? There were still many obstacles to overcome.

Useful enzymes

In 1956 an American scientist, Arthur Kornberg, discovered an **enzyme** called DNA polymerase. It was a copying enzyme. It could copy a single strand of DNA using the building blocks of sugar, phosphate, and bases.

In 1970 Hamilton Smith, in the United States, found a bacterial enzyme. He called it a restriction enzyme. This enzyme was incredibly useful because it could cut DNA, just like a pair of scissors. Today there are many different types of restriction enzyme, each designed to cut DNA at a different point along its length. Using this enzyme, scientists could remove a specific length of DNA, such as the length of DNA that corresponded to a gene.

Both the copying and the restriction enzymes would turn out to be vitally important. They would enable scientists to study pieces of DNA in detail.

British biochemist Professor Frederick Sanger, seen here looking at a model of a DNA molecule, was awarded his second Nobel Prize for Chemistry in 1980.

Working out sequences

In 1977 two research groups, one led by Allan Maxam and Walter Gilbert in the United States and another by Frederick Sanger in the United Kingdom, developed new rapid DNA sequencing techniques. This was only possible because of restriction enzymes.

The Maxam–Gilbert method involved copying the DNA and then carefully cutting it up into fragments. Once they had a lot of DNA fragments, they had to separate them. This was done using **gel electrophoresis**, which moved the fragments different distances according to their length. From the results of the electrophoresis, they could work out the position of each base along the length of DNA.

WHAT IS GEL ELECTROPHORESIS?

Gel electrophoresis is used to separate molecules of different sizes using an electric current. The solution containing the DNA fragments is placed at one end of a sheet of gel. The gel is then placed in a plastic tank and covered with a liquid. DNA fragments have a negative charge. When an electric current is passed through the gel, the DNA fragments move because they are attracted to a positive charge. The gel is made up of a network of tiny threads. Smaller fragments of DNA can pass through the threads more easily than the larger ones. The smaller ones therefore move farther. The DNA is often mixed with a dye. This enables scientists to track its progress across the gel.

In gel electrophoresis, DNA fragments move across a sheet of gel. The DNA is stained with dye so the movement along the gel can be observed.

Working with Genes

Nowadays the process of copying DNA fragments can be carried out in a PCR machine. This automates the process, carefully controlling the temperature of each of the stages.

Scientists needed many copies of **DNA** to analyze. But it was hard for them to get enough DNA. At first they used **bacteria** to produce copies of the DNA, but this took days. Fortunately, in 1983, a scientist named Kary Mullis had a bright idea while driving along a moonlit road one evening in the mountains of California.

Multiplying genes

Kary Mullis' idea was very simple. He thought of using the Polymerase Chain Reaction (PCR). This process made use of DNA polymerase, the **enzyme** that can make copies of DNA.

Once Mullis had explained his idea, others developed it into a process that could be used in laboratories all over the world. Starting with a single strand of DNA, the DNA polymerase makes a copy. Then the process is repeated, and each of the two strands is copied. Each time the process is repeated, the number of DNA samples doubles. Soon there are millions of copies.

WHAT ARE POLYMERASES?

Polymerases are present in most cells. Their job is to copy **genetic** codes, check the copy for errors, and correct any mistakes. This process has been nicknamed "molecular photocopying."

When it was discovered, PCR was described as a breakthrough. It was a quick, easy, and cheap way of producing unlimited copies of any DNA fragment. It works with the tiniest amount of DNA, found in a sample of blood or hair. It even works with DNA fragments that are many thousands of years old. Soon PCR was being used by researchers. It was also being used by forensic scientists to produce evidence in courts of law.

American scientist Tom Brock discovered the bacterium *Thermus aquaticus* in hot springs at Yellowstone Park in Wyoming. This bacterium contains a polymerase that can survive at high temperatures. This makes it suitable for use in PCR machines, where temperatures change rapidly.

How does PCR work?

First the DNA sample is heated. This makes its double helix unwind, and the strands separate. The DNA polymerase produces a copy of each DNA strand. Now there are two double helices. The process is repeated. The DNA is heated again, and the strands separate again. After the next cycle, there are four DNA helices. The process is repeated. Each time, the amount of DNA doubles.

The whole process takes 1 to 3 minutes. In 45 minutes a million or more copies can be produced. However, the sample must not be contaminated with other DNA, as this would also be duplicated. This is particularly important in forensic laboratories, where the evidence produced may be used in court to convict someone of a crime.

Genetic disease

Back in 1902 Archibald Garrod had suggested that some genetic diseases were inherited. Now scientists could search for the **gene** that caused a particular genetic disease. Finding the gene would be the first step toward treating that disease.

Cystic fibrosis is a very serious genetic disease affecting the lungs and digestive system. In 1985 scientists at Toronto's Hospital for Sick Children, in Canada, found the cystic fibrosis gene. It was located on chromosome seven. Then, in 1989, Francis Collins and Lap-chee Tsui, at the University of Michigan, together with John Riordan, at the Hospital for Sick Children, announced that they had worked out the sequence of the 250,000 bases that made up the cystic fibrosis gene.

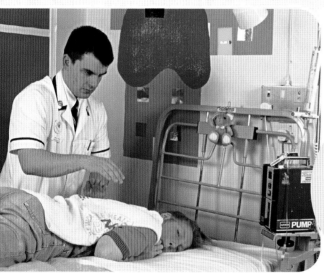

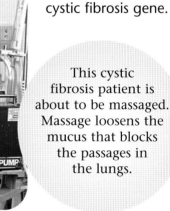

This cystic fibrosis patient is about to be massaged. Massage loosens the mucus that blocks the passages in the lungs.

WHAT IS CYSTIC FIBROSIS?

Cystic fibrosis is the most common inherited disease among **Caucasians** (pale-skinned people). Francis Collins and his team found that 70 percent of cystic fibrosis cases were caused by a single change in one of the DNA codes. This change caused one **amino acid**, phenylalanine, to be left out of the protein. This protein is found in cells lining passages in the lungs. The fault in the protein means that the cells hold in water. Thick mucus then builds up in the passages, which makes breathing difficult.

The discovery of the cystic fibrosis gene enabled scientists to develop a genetic test. The test uses PCR to produce many copies of the person's DNA. This can be checked for the faulty gene. The test can be used to identify carriers of the disease. Carriers are surprisingly common. For example, in the United States about 1 in 25 people are carriers, but only 1 in 2,500 suffer from the disease.

Other tests have since been developed to test for a wide range of genetic diseases. These tests enable would-be parents to check that they are not carriers and ensure that there is no risk of giving birth to a child with a fatal genetic disease.

THAT'S AMAZING!

Some people believe that U.S. president Abraham Lincoln (1809–1865) suffered from a genetic disease called Marfan's Syndrome. The disease weakens the tissue that connects different parts of the body. This can cause the person to develop long limbs, a thin body, and an unusually shaped chest. Lincoln had all these symptoms of Marfan's Syndrome. However, because he died in 1865, nobody can be sure. If a sample of his DNA was available it could be tested for the faulty gene.

This photograph of President Abraham Lincoln, with some army officers, shows that he had the long limbs and thin body that are characteristic of Marfan's Syndrome.

Mapping genes

During the 1990s scientists continued to work out sequences of bases in genes. In 1995 a team of American scientists announced that they had completed a genetic map of the bacterium *Hemophilus influenzae* (*H. influenzae* for short). This was the first complete genetic map of an organism.

This genetic map was different from those that had been produced earlier. The previous maps only gave the relative position of each gene along a chromosome. The new map was far more detailed. It gave the exact position of each gene. It also stated the number of bases from which it was made.

? WHAT IS HEMOPHILUS INFLUENZAE?

H. influenzae is a disease-causing bacterium that infects humans. It can cause childhood ear infections and is linked to meningitis in children. A **vaccine** has recently become available for this bacterium.

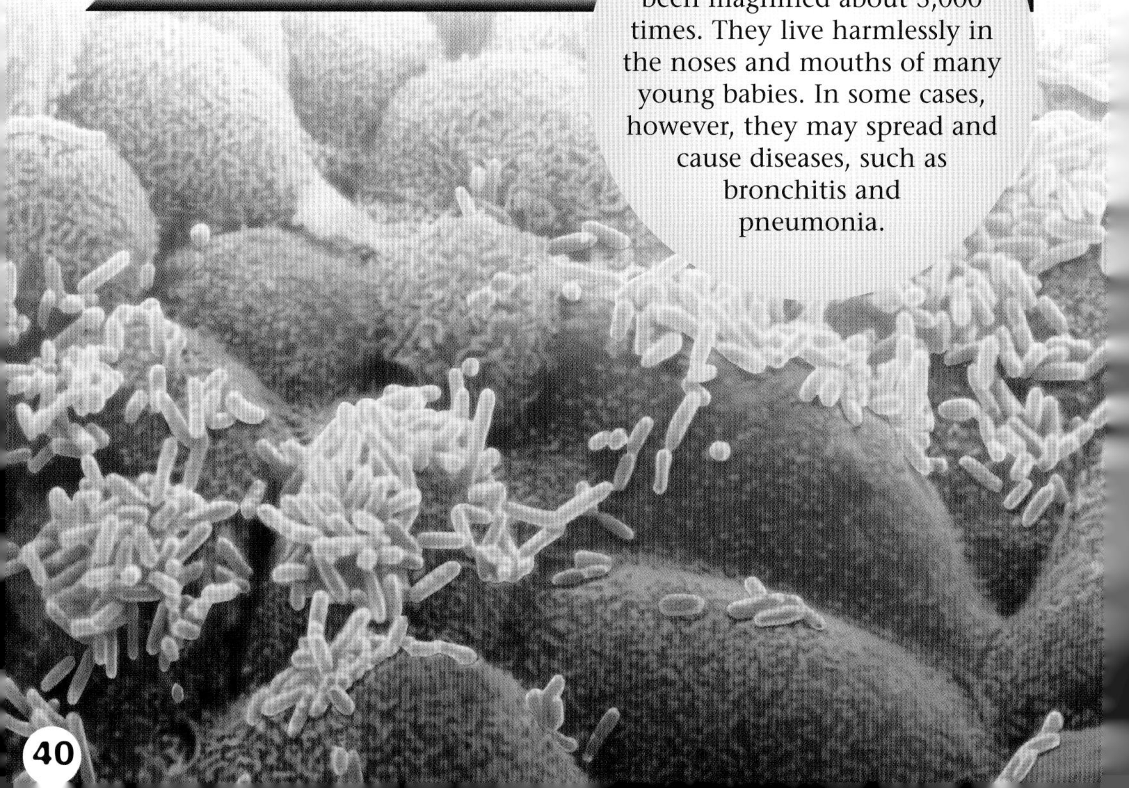

These yellow, rod-shaped *Hemophilus influenzae* bacteria have been magnified about 5,000 times. They live harmlessly in the noses and mouths of many young babies. In some cases, however, they may spread and cause diseases, such as bronchitis and pneumonia.

Better computers

Two developments had enabled scientists to work out this new map. The first was the arrival of more powerful computers during the early 1990s. The second was a new piece of software. Suddenly, because of these developments, genetic sequencing could be done much faster. In addition there was a new approach to sequencing, called "whole **genome** shotgun sequencing."

A new software program, called TIGR Assembler, allowed researchers to sequence large sections of DNA quickly and accurately. It was called shotgun sequencing because the first stage involved breaking apart the organism to get random fragments of DNA. These fragments were sequenced. Then the software program was used to assemble the complete sequence of bases, known as the genome. The fragments are rather like jigsaw puzzle pieces, and the computer has to put the jigsaw together.

This laboratory is at the Massachusetts Institute of Technology. The technicians here are working on mapping genomes. A genome is an organism's total genetic material. Banks of computers are used to analyze the data.

Scientists used this shotgun method to sequence the DNA of the *H. influenzae* genome. In less than a year, the team managed to sequence the 1,830,137 base pairs that made up the 1,749 genes of the bacterium. This would have been impossible just a few years earlier.

A complete genetic map of an organism is useful because it helps scientists work out how a particular organism has changed over time. In the case of disease-causing bacteria, such as *H. influenzae*, this knowledge can help scientists work out which part of the DNA causes harm. This enables them to develop new vaccines to protect people against diseases.

The Human Genome Project

In the 1980s many scientists wanted to work out the human **genome**, the complete base sequence of human **DNA**. The human genome has about 3 billion base pairs. It is much larger than **bacterial** genomes. For instance, *H. influenzae* has just under 2 million base pairs.

Overcoming the problems

During the 1980s some leading research laboratories based in the United States grouped together and applied for money to carry out the work. At this point there was a lot of argument. The project was very expensive and other scientists were worried that their funding would all be used to pay for the human genome studies. Human DNA contains a lot of so-called "junk DNA" that has no specific role. Some people therefore thought there was no need to analyze the entire genome.

However, in the late 1980s developments, such as PCR (see page 37) and computerized sequencing (see page 41), made the project seem more practical. Eventually the government gave its approval.

These automated DNA sequencers analyze lengths of human DNA at the Sanger Center, in the United Kingdom, one of many research centers contributing to the Human Genome Project.

International cooperation

At the same time, similar projects were getting started in countries such as the United Kingdom, Canada, Japan, France, and Italy. They all decided to join forces with the United States. In 1988 a privately funded Human Genome Organization (HUGO) was set up to coordinate the international efforts and collect all the data.

This researcher is preparing human DNA for sequencing. The DNA will be fragmented and then the fragments separated.

The launch of the project

In 1990 what became known as the Human Genome Project was finally launched. The aim of the project was to decode the entire length of human DNA, base by base. At the time it was believed that human DNA contained 40,000–80,000 **genes**. It was hoped that the information gained could be used to help scientists improve medical treatments and develop new drugs. The project was supposed to run for 15 years, finishing in 2005, at an estimated cost of $3 billion.

The Human Genome Project had five main aims:

● to identify all the genes in human DNA

● to determine the sequences of the 3 billion base pairs that make up human DNA

● to store this information in databases

● to improve tools for data analysis

● to look at the moral, legal, and social issues arising from the project.

43

Competing research

The Human Genome Project was funded by governments of countries around the world. This meant that the data could be used by anyone, free of charge. There was also another, privately funded, project sequencing the human genome. This was started by a U.S. company named Celera Genomics, in 1999.

Celera used the shotgun approach (see page 41). Human DNA was broken into fragments and sequenced. Then the company used the largest, most powerful computer in private ownership to assemble the data. Soon Celera Genomics was in a race with the scientists at the Human Genome Project to be the first to publish the human genome.

Celera Genomics used these supercomputers to put together all the sequences from tiny DNA fragments. The computers were very expensive but they enabled the company to complete their human genome map in about a year.

The scientists working on the Human Genome Project used DNA from about ten people, from different ethnic groups, to determine the human genome. Their first draft of the genome was published in 2000. At the same time, Celera Genomics published their own first draft. Since then, the sequences have been checked and corrected, and more detailed studies are now taking place. In 2003 the Human Genome Project was declared complete.

THAT'S AMAZING!

The human genome is a sort of evolutionary history book, revealing changes in humans over time. Some of the oldest sections date back hundreds of millions of years.

What have we learned so far?

The Human Genome Project has discovered that:

- The human genome contains more than 3 billion bases.

- The average gene consists of 3,000 bases, but sizes vary greatly. The largest known human gene is dystrophin, which has 2.4 million bases.

- The total number of genes is estimated to be 30,000. This is much lower than previous estimates of 40,000 to 80,000.

- In humans 99.9 percent of the bases are exactly the same.

- There are 1.4 million places on DNA where the bases vary. A difference in a single base can cause a genetic disease, such as cystic fibrosis.

- We do not know the functions of more than half of the genes that have been discovered.

- Less than 2 percent of the genome codes for proteins.

- About 98 percent of the genome is non-coding DNA. This DNA used to be called junk DNA, but now scientists know that some of it regulates the making of proteins.

- Human chromosomes are numbered from 1 to 23. Chromosome 1 has the most genes (2,968). The Y chromosome (the male sex chromosome) has the fewest (231).

Some of the research centers taking part in the Human Genome Project used automated laboratories and robots to carry out some of the research. This robot arm is loading samples of DNA into plastic plates.

Genetic Fingerprinting

DNA sequencing had some unexpected benefits. In 1984 Alec Jeffreys was working in his laboratory at Leicester University, in the United Kingdom, when he accidentally stumbled upon **genetic fingerprinting**. His discovery soon revolutionized human identification.

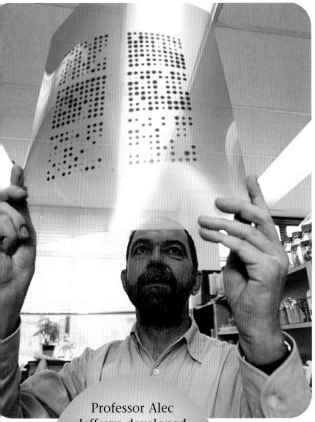

Professor Alec Jeffreys developed the technique of DNA fingerprinting while he was working at Leicester University, in the United Kingdom.

Jeffreys and his team had been studying the differences in DNA between people. At the same time, one member of the team was studying the way **genes** evolve. DNA had been extracted from a sample of gray seal tissue. Jeffreys decided to compare it with human DNA.

The **genetic** fingerprint was produced by chopping up the DNA using restriction **enzymes**. The fragments were separated using **gel electrophoresis**. They were labeled with radioactive markers. A photographic technique was then used to record the position of the radioactive markers. The final result was a series of bars that looked a bit like a barcode on a supermarket product. Two pieces of DNA could be compared by looking at the position of the different bars making up the barcode.

When Jeffreys looked at the first genetic fingerprint it suddenly struck him that this was a really useful technique. It could be used for all sorts of purposes, including identifying individual people, establishing family trees, proving parentage, and researching genetic diseases.

How do scientists make genetic fingerprints?

Genetic or DNA fingerprinting involves five main steps:

1. DNA is removed from the cells or tissues of the body. Only a small amount of tissue is needed. For example, the amount of DNA found at the root of one hair is usually enough.

2. Restriction enzymes are used to cut the DNA into fragments of different lengths. These are sorted according to size, using gel electrophoresis.

3. The pattern of DNA fragments is transferred to a nylon sheet by placing the sheet on the gel.

4. Radioactive probes (types of markers) are added to the nylon sheet to mark the position of the fragments.

5. The sheet is placed on photographic paper and an image is produced. This is the genetic fingerprint.

A scientist uses a magnifying glass to examine a genetic fingerprint. The pattern of bands represents the chemical sequence that forms the genetic code for a section of DNA. The pattern of banding is unique to each person.

Using genetic fingerprints

Each individual has a unique genetic fingerprint, except for identical twins who share theirs. A genetic fingerprint is the same for all of a person's cells, tissues, and organs. It cannot be altered by any known treatment. For this reason, genetic fingerprinting is rapidly becoming one of the main methods of identifying individual human beings.

Genetic fingerprinting enables scientists to identify and copy the tiniest amounts of DNA, even when it is old or damaged. This is very valuable in police investigations. In 1985, within six months of Jeffreys' discovery, genetic fingerprinting had been used in an immigration case to prove the parentage of a Ghanaian boy.

In 1986 genetic fingerprinting was used for the first time in a British criminal case. A suspect was cleared of two rapes and murders, and the real criminal was convicted. In the early 1990s, Jeffreys and his team proved that the remains of a body buried in Brazil were those of the Nazi war criminal Josef Mengele.

Because every organ or tissue of an individual contains the same DNA fingerprint, armed forces have started to collect genetic fingerprints from all their military personnel. The genetic fingerprints can then be used to identify casualties. Genetic fingerprints also have an important role to play in identifying victims of natural disasters.

COULD A DNA DATABASE BE HARMFUL?

Many countries are gathering information from genetic fingerprinting to build huge DNA databases. A national DNA database could be used to help plan health programs and identify criminals. However, there is growing concern that this information could be used in the wrong way by commercial companies or even by governments. There would have to be very strict controls on who had access to the data, and how it was used. For example, banks and insurance companies might use information about a person's risk of disease to refuse them insurance or loans.

Forensics and genetic fingerprints

In a murder investigation, DNA can be obtained from blood found on a murder suspect's clothes or from an object at the crime scene. At first, lawyers claimed that there were too many errors in genetic fingerprinting for it to be used in court. Now, laboratories have to follow certain standards of practice, and genetic fingerprinting is accepted as evidence throughout the world.

Jeffreys' original method used many different markers to create a detailed fingerprint. This takes a great deal of time and money. Much quicker and cheaper commercial methods of genetic fingerprinting have since been developed. These tests are still very accurate and can be used in legal cases.

These forensic scientists are taking samples from blood-stained clothing during a criminal investigation. The genetic fingerprint from the blood sample will then be compared with DNA from the victim and any suspects.

Family Trees

People vary in size, height, skin color, hair type, blood group, and much more. However, our **DNA** is not that different. In fact more than 99 percent of our DNA is identical. Only a fraction of 1 percent of our DNA is responsible for most of the differences. It is this fraction that interests scientists.

Human DNA is helping scientists to find out about the evolution of humans. Scientists used to have to rely on evidence from human fossils in order to piece together the evolutionary path. However, over the last 15 years, new techniques have been developed as a result of advances in PCR (see page 37), the Human Genome Project, and **genetic fingerprinting**.

Dr. Bryan Sykes, of Oxford University, UK, studied DNA taken from the "Ice Man," whose frozen body was discovered in Austria in 1991. Sykes showed that millions of present-day Europeans are genetically related to the Ice Man, who lived about 5,000 years ago.

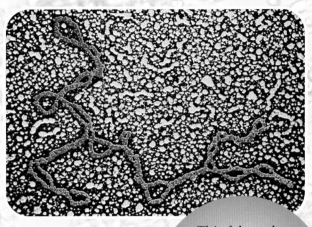

Mitochondria are the powerhouses of cells. Their role is to break down molecules and generate energy that can be used by other parts of the cell.

This false color photograph, taken with a powerful microscope, shows a strand of mitochondrial DNA (in orange), magnified about 300,000 times.

Comparing DNA

Scientists can compare the DNA of humans of all races in order to build family trees. Changes regularly occur in human DNA, and these changes are passed on to the next generation. It is these changes that make each one of us unique. By analyzing the differences, scientists can work out how closely we are related.

The DNA that is used to build family trees does not come from the cell **nucleus**. Instead, scientists use DNA from a mitochondrion, a tiny structure in the gel surrounding the nucleus. **Mitochondria** contain a short length of DNA (about 16,000 to 17,000 base pairs), with 37 **genes** that are involved with producing proteins. Mitochondrial DNA undergoes changes just like other DNA, but at a faster rate. More changes make comparisons easier, as there are more differences.

Human evolution

In 1987 Rebecca Cann, Mark Stoneking, and Allan Wilson, based in California, published the results of their analysis of mitochondrial DNA in different human races. They concluded that we all have a common ancestor who was probably living in Africa about 200,000 years ago. Since that time people have moved from Africa and spread across the world.

Inheritance Today and Tomorrow

Our knowledge of inheritance has progressed a long way in a relatively short time. But new discoveries are still being made, and the future could be just as interesting as the past.

The Human Genome Project (see pages 42–45) has revealed that humans have only twice as many **genes** as a fly. Yet the human body is far more complex than a fly's body. If it is not the number of genes that produces a complex body, what could it be?

Scientists think the answer may lie in the role played by certain genes. Many genes are involved in managing or controlling other genes. There may be managers of managers, too! Scientists are trying to work out how all these managers work together.

This cystic fibrosis patient is taking part in trials of a drug treatment called Alpha 1. This is a protein that is in healthy human lungs, but only present in low amounts in people with cystic fibrosis.

Gene therapy

Scientists have identified several genes that cause disease. Some of these diseases may be cured, or prevented, by stopping the gene from working. This is called gene therapy.

Gene therapy was first proposed more than 30 years ago. It involves inserting a length of **DNA** into a cell to replace the faulty DNA. The main problem has been getting the DNA into the cell. Different methods, such as using **viruses** to carry the DNA, have been tried.

In 1993 cystic fibrosis became the first **genetic** disease to be treated with gene therapy. Scientists had altered a common cold virus to act as a delivery vehicle. The cold virus carried the normal DNA into the damaged cells of the lung passages. Patients can now use a nasal spray to get the viral particles into their lungs. This method has had some success. However, the lung cells are constantly replaced with faulty cells, so the treatment has to be repeated. In the future scientists hope to find a permanent treatment.

Switching off genes

Scientists are also investigating ways of switching off genes in order to treat Huntington's disease. In 2003 Dr. Beverley Davidson, of the University of Iowa, stopped a progressive brain disease in mice by using a genetic technique called **RNA** interference. This blocks the RNA produced by the faulty DNA, shutting down the faulty gene.

Doctors are very excited about this development. There is now a genetic test available to identify Huntington's sufferers before the symptoms appear. This means that the new treatment could be given early on, before too much damage is done to the brain. The first clinical trials on patients may begin before 2010.

TALKING SCIENCE

"When I first heard of this work, it just took my breath away. It's everything you ever wanted to hear and more."
Nancy Wexler, Columbia University, talking about gene therapy for Huntington's disease

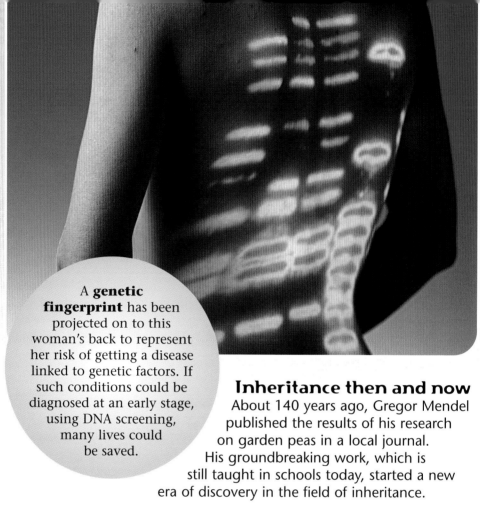

A **genetic fingerprint** has been projected on to this woman's back to represent her risk of getting a disease linked to genetic factors. If such conditions could be diagnosed at an early stage, using DNA screening, many lives could be saved.

Inheritance then and now

About 140 years ago, Gregor Mendel published the results of his research on garden peas in a local journal. His groundbreaking work, which is still taught in schools today, started a new era of discovery in the field of inheritance.

A century ago, scientists had just discovered chromosomes in cells. Then Thomas Hunt Morgan made a major contribution to the understanding of inheritance with his work on fruit flies. From this point on, research focused on genes, and the chemical from which genes are made: DNA.

The human genome and medicine

Now scientists have put together the entire human **genome**, and this knowledge is having an impact on genetic diseases. Genes have been linked with certain types of breast cancer, muscle disease, brain disorders, deafness, and blindness, among others.

Scientists are also finding the DNA sequences that are linked to common health problems, such as heart and lung disease, diabetes, arthritis, and many cancers. This knowledge may lead to the development of effective new gene therapies.

Another surprising discovery arising from the Human Genome Project was the number of single base changes in the code. Some of these changes in DNA can be inherited. But they can also be accumulated during a person's lifetime, because of environmental pollution, food, drugs, and so on. Most of these changes are not passed on, because they affect the body cells and not the sex cells. For example, a woman may get skin cancer from getting sunburned. However, the damage to the DNA in her skin cells will not be inherited by her children. The only DNA changes that she can pass on are changes that lead to alterations in her egg cells.

Other benefits

There have been other benefits from research into inheritance. **Enzymes**, such as DNA polymerase and restriction enzymes, have made it possible to cut out lengths of DNA from one organism and paste them into another. This is called **genetic engineering**. Over the last 30 years, many different genetically engineered organisms have been produced in order to manufacture foods and medicines.

In the past, scientists had to study one or two genes at a time. Now they can study large groups of genes. For instance, they can investigate all the genes in a particular type of tissue or organ. They can also start to unravel the mystery of how thousands of genes work together to control the complex functioning of the human body. This is the challenge facing today's scientists, and it may take many decades before they find the answers to their questions.

A thick layer of smog lies over Shenyang, one of China's leading industrial cities. Over time, chemical pollutants in the air can alter a person's DNA, leading to disease.

Timeline

1866   Gregor Mendel publishes his investigations into inheritance of pea plants.

1890   Theodor Boveri suggests that chromosomes are involved with inheritance.

1900   Mendel's work is rediscovered by three scientists: Hugo DeVries, Erich Von Tschermak, and Carl Correns.

1900   Walter Sutton observes chromosomes in grasshopper cells.

1902   Archibald Garrod discovers that some diseases must be inherited.

1903   Sutton and Boveri, working independently, suggest that each egg or sperm cell contains only one of each chromosome pair.

1905   Edmund Beecher Wilson and Nettie Stevens, working independently, propose that certain chromosomes determine sex. They show that a single Y chromosome determines maleness, and two copies of the X chromosome determine femaleness.

1909   Wilhelm Johannsen uses the term "gene" to describe the carrier of heredity, "genotype" to describe an organism's genetic make-up, and "phenotype" to describe an organism's outward appearance.

1910   Thomas Hunt Morgan proves that genes are carried on chromosomes. He also shows that some characteristics are carried on the sex chromosome.

1911   Alfred Henry Sturtevant maps the genes on the fruit fly's sex chromosomes.

1912   Sir William Henry Bragg and his son discover that X-rays can be used to study the molecular structure of simple crystals, such as salt.

1926   Morgan publishes *The Theory of the Gene*.

1944   Oswald Avery, Colin MacLeod, and Maclyn McCarty use bacteria to show that DNA is the hereditary material.

1949   Erwin Chargaff finds that the amounts of adenine and thymine in DNA are about the same, as are the amounts of guanine and cytosine.

1953 James Watson and Francis Crick propose that the DNA molecule is a double-stranded helix.

1963–1966 Marshall Nirenberg and Heinrich Matthaei work out the genetic code.

1977 DNA from a virus is sequenced for the first time by Frederick Sanger, Walter Gilbert, and Allan Maxam, working independently.

1983 Kary Mullis discovers the Polymerase Chain Reaction (PCR), enabling lengths of DNA to be multiplied.

1984 Alec Jeffreys produces the first genetic fingerprint.

1987 Rebecca Cann, Mark Stoneking, and Allan Wilson analyze mitochondrial DNA in different human races. They declare that humans have a common ancestor who lived 200,000 years ago.

1989 The first human gene is sequenced by Francis Collins and Lap-chee Tsui. It is the gene that causes cystic fibrosis.

1990 The Human Genome Project is launched.

1993 Cystic fibrosis becomes the first genetic disease to be treated using gene therapy.

1995 The genome of *H. influenzae* is sequenced. This is the first complete genome of an organism.

2000 First draft sequences of the human genome are released at the same time by the Human Genome Project and Celera Genomics.

2003 The Human Genome Project is successfully completed on April 14.

Biographies

These are some of the leading scientists in the story of inheritance.

Theodor Boveri (1862–1915)

Theodor Boveri was born in Bamberg, Germany. He went to the University of Munich, where he gained his doctorate. In 1887 he qualified as a university lecturer in zoology and comparative anatomy. Then, in 1893, he was appointed Professor of Zoology and Comparative Anatomy at the University of Würzburg. His research on roundworms and sea urchins showed that it was necessary to have all chromosomes present for an embryo to develop correctly.

Francis Crick (1916–2004)

Francis Crick was born in Northampton, in the United Kingdom, and studied physics at University College, London. After World War II (1939–1945), he went to Cambridge to study biology. In 1951 he met James Watson and in 1953 they proposed the structure of DNA. Crick later did research in biochemistry and genetics. He was made a Fellow of the Royal Society in 1959. In 1960 he was presented with an Albert Lasker Award for Basic Medical Research. In 1962 he was awarded the Nobel Prize for Physiology or Medicine, shared jointly with Watson and Maurice Wilkins.

Gregor Mendel (1822–1884)

Gregor Mendel was born in what is now called the Czech Republic. From 1851 to 1853, he studied zoology, botany, chemistry, and physics at the University of Vienna. He then became a monk at an Augustinian monastery in Brno. There, in 1856, he began his experiments with pea plants. He became a member of the Zoological–Botanical Society of Austria and published two scientific papers in 1853 and 1854. After spending eight years carrying out research, Mendel gave two long lectures that were published in 1866 as "Experiments in Plant Hybridization." This paper set out his laws of inheritance.

Thomas Hunt Morgan (1866–1945)

Thomas Hunt Morgan was born in Lexington, Kentucky. He went to the University of Kentucky, where he received a science degree. In 1890, he obtained a Ph.D at Johns Hopkins University. The following year, he was awarded the Adam Bruce Fellowship. This enabled him to spend a year in Europe. There he met renowned embryologist Hans Driesch who influenced some of his early research. In 1891 he became Associate Professor of Biology at Bryn Mawr College for Women, where he stayed until 1904.

He then moved to Columbia University to become Professor of Experimental Zoology. Over the next 24 years, he carried out his groundbreaking research on the fruit fly. In 1928, he was appointed Professor of Biology and Director of the G. Kerckhoff Laboratories at the California Institute of Technology. His publications include *Evolution and Genetics* (1925) and *The Theory of the Gene* (1926). He received the Nobel Prize for Physiology or Medicine in 1933.

Nettie Stevens (1861–1912)

Nettie Stevens was born in Vermont, and went to school in Westfield, Massachusetts. She gained two degrees from Stanford University and received a Ph.D. from Bryn Mawr College for Women in 1903. She then stayed on at Bryn Mawr, as a researcher. In 1904 Stevens wrote an important paper on regeneration with Thomas Hunt Morgan. Her work on regeneration led her to study differentiation in embryos and then to study chromosomes. In 1905, she announced her finding that the X and Y chromosomes were responsible for determining sex. The same year Edmund Beecher Wilson independently announced that he had made the same discovery. Stevens continued her research on chromosomes in various insects. She died in Baltimore, Maryland in 1912.

James Watson

James Watson was born in Chicago, Illinois, and went to the University of Chicago when he was only 15 years old. He received his Bachelor of Science degree in zoology four years later, and went on to gain a doctorate in zoology at Indiana University. He was given a research post at the University of Copenhagen, in Denmark, where he heard about the biomolecular research that was being carried out at Cambridge University in the United Kingdom. He joined Francis Crick in Cambridge in 1951. They published their paper on the structure of DNA in 1953. In 1962 Watson, Crick, and Maurice Wilkins were awarded the Nobel Prize for Physiology or Medicine. In 1968 Watson published his account of the DNA discovery, *The Double Helix*, which became an international bestseller. He moved back to the United States, where he has become a respected figure in genetics research. During the late 1980s and early 1990s, he was head of the Office of Human Genome Research at the U.S. National Institutes of Health.

Glossary

allele form of a gene

amino acid one of the basic chemicals of life. Proteins are made up of different combinations of about 20 amino acids.

autosome chromosome that is not linked to sex

bacteria single-celled organisms that are found everywhere, some of which cause disease

Caucasian member of one of the human races that has light-colored skin

dominant (of a characteristic) showing itself even when there is only one copy of the allele

dominant allele form of a gene that will always show itself. For example, brown eyes are caused by a dominant allele, and a person carrying one allele for brown will have brown eyes. The brown masks the color blue.

DNA (deoxyribonucleic acid) molecule that carries the body's genetic information and makes up chromosomes

embryo term given to the new individual that forms when an egg is fertilized

enzyme type of protein that speeds up chemical reactions

factor term used by Gregor Mendel for what is now called a gene

gamete sex cell (e.g. egg or sperm) of a living thing

gel electrophoresis method used to separate molecules of different sizes, using an electric current passed through gel

gene unit of inheritance that is passed on from parent to offspring, made up of a length of DNA

genetic inherited; having to do with information that is passed from parents to offspring through genes

genetic engineering altering DNA in a laboratory

genetic fingerprinting process of chopping DNA into fragments, using restriction enzymes. The fragments are separated, then labeled with radioactive markers.

genome complete sequence of bases in an organism's DNA

heterozygous having two alleles that are different

homozygous having two alleles that are the same

larva growing stage in an insect's life cycle, between the egg and the winged form

meiosis type of cell division, also known as a reduction division, in which the number of chromosomes is halved to form the gametes

mitochondria structures found in cells, which are responsible for producing energy that is used by the rest of the cell

mitosis type of cell division used in growth, in which a cell divides into two identical daughter cells

nucleus central part of the cell, which contains DNA and controls many cell functions

pupate (of an insect) start to change from a larva into a winged adult

recessive (of a characteristic) only showing itself when two alleles of the gene are present

recessive allele form of a gene that can be masked by a dominant allele. For a recessive characteristic to show itself, two alleles must be present.

RNA (ribonucleic acid) type of nucleic acid, related to DNA

stamen male part of a flower, consisting of the anther, which contains the pollen, and supporting structures

stigma section of the female part of a flower, containing the eggs, from which seeds will grow. When pollen lands on the stigma, the pollen grains grow tiny tubes that enable them to insert their genes into the eggs.

vaccine substance containing dead or weakened microbes that is introduced into the body in order to provide protection against a disease

virus disease-causing microbe that is many times smaller than a bacterium

Further Resources

If you have enjoyed this book and want to find out more, you can look at the following books and websites.

Books

Bankston, John. *Gregor Mendel and the Discovery of the Gene.* Hockessin, DE: Mitchell Lane Publishers, 2004.

Claybourne, Anna. *Usborne Internet-Linked Introduction to Genes and DNA.* Tulsa: Usborne, 2003.

Seiple, Samantha and Todd. *Mutants, Clones, and Killer Corn: Unlocking the Secrets of Biotechnology.* Minneapolis: Learner Publishing Group, 2005.

Solway, Andrew. *Life Science In Depth: Inheritance and Selection.* Chicago: Heinemann Library, 2006.

Websites

Human Genome Project information
www.ornl.gov/sci/ techresources/Human_Genome /home.shtml
This website explains the aims, history, and progress of the Human Genome Project.

How Stuff Works
science.howstuffworks.com/ gene-pool1.htm
This website offers very clear explanations and diagrams of DNA and inheritance.

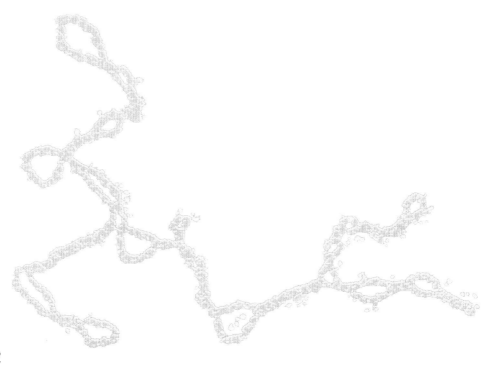

Index

Index